MARJORIE HARRIS

Favorite Flowering Shrubs

Photographs by PADDY WALES

HarperCollins*PublishersLtd*

 For information address HarperCollins Publishers Ltd, Suite 2900, Hazelton Lanes, 55 Avenue Road, Toronto, Canada M5R 3L2. Distributed in the United States by HarperCollins International, 10 East 53rd Street, New York, NY, 10022.

First Edition

Canadian Cataloguing in Publication Data

Harris, Marjorie
Marjorie Harris favorite flowering shrubs

ISBN 0-00-255394-5 (bound)
ISBN 0-00-638025-5 (pbk.)

1. Flowering shrubs — Canada. I. Title. II. Title: Favorite flowering shrubs.

SB435.H377 1994 635.9'76 C94-930696-7

94 95 96 97 98 99 ❖ RRD 10 9 8 7 6 5 4 3 2 1

Printed and bound in Mexico

Design: Andrew Smith
Page layout and composition: Joseph Gisini, Andrew Smith Graphics, Inc.
Editing: Barbara Schon

ACKNOWLEDGEMENTS

A thank you to Jacqueline Rogers for her splendid research, to Barbara Schon for the difficult task of editing, Andrew Smith for his brilliant design and Maya Mavjee for providing the gentle cheerleading. My admiration for Paddy Wales is boundless. She found the plants, and she got these wonderful gardeners to let her photograph them to have in our books: Ann Buffam, Capilano Suspension Bridge, Pam Frost, Kathy Leishman, Maple Leaf Garden Centre, Helen and Don Nesbitt, Phoebe Noble, Glen Patterson, Dick and Barbara Phillips, Joan Rich, Susan Ryley, University of British Columbia Botanical Gardens, VanDusen Botanical Garden and Stewart and Pauline Webber.

COVER: *Magnolia quinquepeta*

Contents

Flowering Shrubs

Flowering shrubs are the backbone of the garden. Even one shrub in the perfect spot can make your garden extraordinary. Without shrubs, a garden is a sorry place where it is almost impossible to bring pleasure through all four seasons—something all gardeners aim for.

And though shrubs may ultimately be less demanding than annuals or perennials, proper attention is crucial right from the start. Most shrubs need faithful watering during the first year, but with suitable mulching will survive quite nicely on nature's bounty after that. Since I garden organically, I don't add manufactured chemicals and never have. Feed the soil rather than the plant, and you'll take the guesswork out of gardening.

There are thousands of shrubs, hundreds of them available at nurseries. To make my choices for this book I used two criteria: I've grown them myself, or they are plants that I love but can't grow in my climate. The selection was further tempered by what was available when Paddy Wales was taking her excellent photographs. She had limited time—three seasons to do all the work in this series—and was often up against inclement weather in her search for the best, the most beautiful form of each shrub. And she pulled it off. We missed a few: I love lilacs, for instance, but we could never get the right shot. And there wasn't room for exochorda, stewartia, callicarpa or many other splendid plants. So we pared our list down to some of my favorites.

I also wanted plants that would be available in as many zones as possible. For instance, if you can't use the one photographed and you like it, look for another one in the same genus for your area, or a plant close to the one I love that will give the same satisfaction. Since nothing grows alone in nature, I've also included some of my favorite combinations.

HOW TO USE SHRUBS

I use shrubs as screening plants; for example, I have a buddleia in front of the deck where we sit out in the evening. It's beautiful to look at in itself and it also serves as a theatrical scrim through which we look at the other plants in the garden. The fewer specifics shown on first viewing, the more mysterious the whole becomes, and the more ravishing as you travel through it.

Use shrubs to guide the viewer through the garden, section by section, or to shroud something ugly. Mass them together to form an interesting border just by their very contrasts in leaf shape and size.

Use shrubs the same way you do perennials—mix them up with all sorts of other plants rather than keeping them isolated or, most boring of all, as foundation plants.

Every shrub has a face, the angle at which it looks best. In the nursery before I make my choice, I move them around and around, and then repeat this again when I get it into the garden. Having a handsome profile is important for any plant, and even more so for such vital plants as shrubs.

Obviously there are some shrubs that simply won't grow in my region, which is zone 6. But one of the amazing things about gardening is that you can take universal principles and apply them to gardening anywhere. The first and most important of these principles is knowing as much as possible about your own microclimate.

YOUR MICROCLIMATE

First of all, find out what your general hardiness zone is. The zones used in this book follow the U.S. Department of Agriculture designation for the limits of hardiness (the lowest possible temperature at which a plant can survive). These temperatures are on page 9. Find out how low the average temperature in your area plunges each year and you have a general idea of the hardiness zone. For some shrubs there is contradiction and controversy over hardiness zones. When there is no general agreement, I've used the

Buddleia davidii 'Pink Delight'
PHOTOGRAPHED IN: University of British Colombia Botanical Gardens

recommendations of John J. Sabuco and Michael Dirr (see Bibliography).

Hardiness zones don't take into account such important elements as altitude, prevailing winds or whether you live in a frost pocket. Frost settles into valleys or even depressions, and these areas will likely suffer more, deeper and earlier frost hits. A fence on a hillside above the garden will block some of the cold air pouring into a depression, but if it is on the downhill side below the garden, it will keep the cold air in place.

Even all this information won't tell you what's going on in your own backyard. Local weather conditions and the light and soil in your garden make up your microclimate, and it may be different from one down the road or ten blocks away.

And that's important to keep in mind when choosing plants. Depending on the microclimate in your garden, you may be able to push the limits of the zone you live in. If you live in zone 4, but your garden is surrounded by tall buildings or big trees that serve as windbreaks, or you have lots of raised beds, you may be successful with plants that would never be considered hardy to your area.

To get even more detailed information about your garden, keep track of the area temperature, garden temperature and dates of the first and last frost in general then in your own garden.

SOME GENERAL WEATHER INFORMATION

- Temperatures taken at ground level are lower than temperatures a few feet above the ground. You can buy thermometers that register both. Record the general temperature to see how it varies from that of your garden.
- A clear night when the temperatures hit 40–45°F (4.5–7°C) will warn you that frost is on its way in the near future.
- Clouds and wind lessen the chance of frost. Hang a piece of plastic ribbon in a tree; as long as it's moving you probably won't have frost.
- East of the Rockies, a westerly wind means fair weather.
- West of the Rockies, a westerly wind means rain or snow.

PLANTING

The safest way to get your shrubs started on a long and healthy life is to use good planting techniques. Try the following:

- Dig a hole exactly the same depth as the root system and about five times as wide. Do not change the soil by adding any amenders such as compost or manure. They will mollycoddle a plant. The roots of most plants grow outwards, so when they get beyond the area you've fertilized they'll be in alien soil and likely react badly. Start them in the soil they have to live with.
- Check the drainage. Some plants can cope with wet feet, but most prefer good drainage. After you've dug the hole, pour in a few gallons of water

Kalmia latifolia
PHOTOGRAPHED IN: University of British Colombia Botanical Gardens

and see how long it takes to drain away. If water is still sitting there hours later, that's what your plant will have to contend with. If that is the problem, dig out a very large, deep area. Add gravel to the bottom and replace the natural soil. Then plant.

❧ Do all the amending to the surface. Add compost, manure and any other fertilizers to the area around the plant. This is called top dressing. The nutrients will naturally be pulled into the soil by worms and microbic denizens to feed the plant as required. Add mulches to the surface to help hold in moisture, protect the root system and keep down weeds.

❧ If your soil doesn't have the right pH (acidity or alkalinity) for the plant you've chosen, either dig out a huge area and completely change the soil or use raised beds and do the same. It's easier to change the plant than the soil, so experiment with small plants first to see how they like the area. If they die, it isn't a big expense.

❧ The lower the pH number, the higher the acidity; neutral is about 7 on a scale of 1 to 14. It's difficult to change the pH of a soil but it can be done. To raise the alkalinity, add lime. Sulphur will raise acidity, but never use aluminum sulphate—it is poison to plants. It's safer to use pine needles, sphagnum peat moss and oak leaf mold. They will raise acidity very slowly.

❧ There is a theory that mulch around a plant "pulls" the nitrogen out of the soil in the process of decay. This doesn't happen in the forest and doesn't seem likely to happen in your backyard. A greater risk is that if the mulch is dry it may draw too much water out of the soil. Before mulching, water the soil and allow the water to drain away. If a plant at first reacts badly to mulching, it may only be adjusting. Be patient.

CREATIVE PRUNING

Russell Page, my all-time hero of garden design, talked lyrically about only really understanding pruning when he got underneath a *Magnolia soulangiana* and looked up: "I began to prune... starting rather gingerly, with a twig here and there. As I worked I realized that I was working with space, carving the empty air into volumes caught in the angles of branch crossing branch and held by leafy sprays; and that here in the circumference of a small tree lay the meaning of a whole relationship between art and nature."

Creative pruning is just that, making a beautiful shrub look even better, and keeping it healthy and happy. The listings describe specific care for each shrub, but there are a few general principles to keep in mind when you are pruning. If you are a complete novice, practice on a shrub you don't care about or want to scrap. Who knows—you may end up with a shrub you can't part with.

Though I can tell you what to look for, close observation is your own best guide to creating a well-shaped shrub. Be sure to follow these rules:

- Never cut into the collar of the branch where it is attached to the trunk. If you prune properly, a callus will form around the wound and protect the plant from disease and bugs. If you cut too close to the trunk, a callus won't form and water shoots (ugly vertical branches) will pop out everywhere along the stub.
- On fruit trees, buds for flowers and fruit form on short spurs. Avoid these when you are pruning.
- As a general rule, take out one-eighth to one-quarter of the leaf-bearing area.
- Thinning will reduce the bulk of a shrub: take out the branch right at the spot where it starts. This will make a shrub more slender since these branches tend not to return.
- Selective heading will cut a shrub back to the size you want to keep it: cut back the branch to a side branch about half the size of the branch being removed. Anything that's really skinny can be taken back to a bud.
- Non-selective heading will make any shrub look bushier: cut back any branch at any point to stimulate new growth.
- Flowers seem to grow best on wood that's from one to four years old.

Remove one-third of the oldest stems each year to renew the plant.
☙ Canes, such as forsythia, hydrangea, mahonia, kerria and weigela, renew themselves by growing canes or new branches from the base of the plant. Take out all the deadwood and any canes that flop about and make the shrub look messy. Cut right back to the ground. Remove cross-overs or anything rubbing against each other. Do this every year. If the plant gets sloppy looking, chop it right back to the ground and leave it alone for a couple of years.
☙ Mound-shaped shrubs such as escallonia, azalea, ilex, spirea and abelia have soft stems and small leaves. Cut out the longest branches right to the interior of the shrub. This will disguise the cut. The idea is to make the shrub look tidier and keep its size.
☙ Tree-like shrubs such as pieris, rhododendron, witch hazel, camellia, deciduous azalea and cotoneaster have woodier branches that divide many times. Work from the bottom towards the outside of the plant. Keep moving around the shrub so you don't get overzealous on one side. Take out deadwood to start with—this could, in fact, be all you need to remove. Then cut out the usual cross branches, anything that rubs together. Open up the centre. Remove branches that brush the ground and suckers.
☙ Roses should be shortened and thinned while the plants are dormant. Shortening means cutting back to the desired height; during the next season canes will reach their natural length. Thinning removes weak, crossing branches and creates more lateral branches. Cut just above, but not too close to, a bud. The new branch will come from that bud, and it will grow in whatever direction the bud faces—cut near outward-facing buds to get an outward-facing branch.
☙ Evergreens are best pruned in spring.
☙ Usually only hedges and topiary are sheared.

ZONE CHART		
Zone 1	below –50°F	(below –45°C)
Zone 2	–50 to –40°F	(–45 to –40°C)
Zone 3	–40 to –30°F	(–40 to –35°C)
Zone 4	–30 to –20°F	(–35 to –30°C)
Zone 5	–20 to –10°F	(–30 to –23°C)
Zone 6	–10 to 0°F	(–23 to –18°C)
Zone 7	0 to 10°F	(–18 to –12°C)
Zone 8	10 to 20°F	(–12 to –7°C)
Zone 9	20 to 30°F	(–7 to –1°C)
Zone 10	30 to 40°F	(–1 to 4°C)

JAPANESE MAPLE

Acer palmatum

'Dissectum Atropurpureum'

FAMILY NAME: *Aceraceae* / ZONE: 5 or 6, depending on species
PHOTOGRAPHED IN THE GARDENS OF: Glen Patterson and Kathy Leishman

Many years ago I bought an *Acer palmatum* 'Dissectum Atropurpureum'. At the time I didn't know the botanical name of this Japanese maple and it seemed terribly expensive ($20). But I was drawn inexorably to its elegant shape. The boughs didn't so much branch as spill away from the trunk of the tree in a graceful arch. The deeply cut leaves of this particular maple are a sensuous burgundy that start to turn green from the centre of each leaf in autumn and finally end up in a riot of brilliant scarlet and yellow. Maybe it's my imagination, but the leaves all seem to drop on the same day, making a bed of lacy litter beneath the vaulted branches. I wouldn't touch this picture for anything.

I plunked my little shrub into a central spot in the garden where it's remained to this day. Though I've redesigned the garden several times—and I move plants constantly—I always leave that Japanese maple in its place of honor.

And this plant deserves such attention. *Acer palmatum* comes in every imaginable height, from a little guy like mine, which has never reached more than 2 feet (60 cm) in 10 years, to shrubs the size of small trees (over 21 feet/7 m) up to giants of 50 feet (15 m). The foliage is astonishing because it comes in hugely diverse colors ranging from pink to green to yellow, variegations and thrilling reds in all tones.

The elegance of shape makes almost any Japanese maple truly flexible. It's a splendid shrub to use as a specimen, to define the border of a woodland or in the central part of a mixed border. About the only place this plant is in danger of being a cliché is stuck smack dab in the middle of a front lawn all by itself. It comes from Japan, China and Korea, where it fits into the undulating landscape with other plants. It seems only fair to treat it similarly here.

There are hundreds of cultivars, enough to suit almost any purpose in the garden. Some have green leaves with bright red seeds; or red leaves with bright green seeds; spring color as brilliant as any azalea; incredible fall

Acer palmatum 'Dissectum Atropurpureum'

color; or useful forms and shapes. A flexible plant indeed.

Acers may look finicky or delicate but they aren't—if they are properly sited and carefully planted. Once that's accomplished they need little attention. These are very expensive plants, and it's risky moving them about. In other words, do the fussing before you dig.

The plant thrives in zones 5 and warmer, but find out the needs of your specific acer. Some can take the sun, but others will "bronze out", or change color, when exposed to midday sun. This is especially true of the deeper red forms.

Then there is the question of what to put under or near a Japanese maple. The almost neon brilliance of the foliage in the fall must be considered, and it's important to give it enough breathing room. But don't be callow about placement. One of the most stunning uses I've seen is in the garden of a collector. But her garden doesn't shout "Collector Here". Dozens and dozens of Japanese maples of all sizes and colors are worked into perennial and foliage borders with such grace that only the overall harmony of the garden is evident. They are surrounded by broad-leaved evergreens such as rhododendrons, pieris and mahonias, plus a combination of low and spiky perennials.

Under them I like a low background of *Asarum europaeum*, European ginger, but almost any deep green ground cover will do just as well.

PRUNING TIPS

❧ Upright forms: They need occasional pruning of the inside branches to let some light in, but only if absolutely necessary. This will also reveal the wonderful twisted form of the trunk. In winter, when it's dormant, prune the tips to shape the tree.

❧ Weepers: These forms seldom need anything done to them. Remove any dead branches, and those that sprout from the centre of the tree. Keep it shapely.

Acer palmatum 'Dissectum Atropurpureum'

PLANTING & MAINTENANCE TIPS

❧ Zone of tolerance: to warm parts of zone 5. If you have a shrub grown on its own roots (and you must check this with the grower), it will be hardy to –30°F (–35°C); on grafted root stock to –20°F (–30°C).

❧ The soil isn't as important as placement: a half day of full sun at the most. Avoid putting the deep red forms in sites with intense midday sun. And keep them out of windy blasts. Prepare the soil in the usual way (see page 6–8) and make sure you plant it at exactly the same depth as at the nursery.

❧ It's important to stop fertilizing after about the middle of August. You don't want to encourage new growth that won't harden off before the first frost sets in.

❧ The rule of thumb is that these plants should never, ever be allowed to dry out when they are young. Real mavens keep them mulched even after they're well established.

OTHER SPECIES & HYBRIDS

Japanese maples found in nurseries are mostly cultivars developed from *A. palmatum* (the most common) or *A. japonicum*. The two species divide, respectively, into upright and weeping forms. *A. palmatum*, which means resembling a hand, has 1½" – 2" (4 cm – 5 cm) lobes radiating from the centre of the leaf; the tree grows to 15' – 30' (4.5 m – 9 m). *A. p.* 'Dissectum' is a weeping form. *A. japonicum* is a weeper growing laterally rather than upright. Most are slow growing, up to 3 feet (1 m) high and sideways in the shape of a mound. And the leaves of these cultivars are larger—3"– 4½" (7.5 cm – 11 cm)—in pale gold to dark green, from deeply divided to an almost circular shape.

Acer palmatum 'Hogyoku', new growth is bronze, bright green in summer, orange in fall. Grows to 10 feet (3 m). 'Oshio-beni', red-orange leaves; 'Sangokaku' (Sen Kaki), coral bark maple, is a heart–stopper. The stems really are a strong coral color, and the leaves are almost apricot in fall. With winter protection, it should be hardy in zone 6.

Other members of the *Acer palmatum* family to look for: *A. p.* 'Atropurpureum'; *A. p.* var. *heptalobum*; *A. p. h.* 'Osakazuki'; *A. p.* 'Linearilobum'; also *A. p.* 'Versicolor'; *A. p.* 'Pink Edge'. They are generally hardy in warm parts of zone 5 and definitely in zone 6.

* If you cannot grow any of these Japanese maples try a wonderful substitute: *Betula albo-sinensis* 'Trost's Dwarf', which has the cut leaf of *A. p.* 'Dissectum' but is hardy in zone 5.

BUTTERFLY BUSH

Buddleia davidii var. *nanhoensis* 'Petite Plum'

FAMILY NAME: *Loganiaceae* / ZONE: 5 to 9
PHOTOGRAPHED IN THE GARDENS OF: Kathy Leishman and University of British Columbia Botanical Gardens

Some plants hit the top of my all-time favorites list and then plummet for one reason or another. But buddleias will never fade from favor, will never be anything but perfection. The butterfly bush has so many virtues it's hard to imagine a garden without it. It really does attract butterflies, and there is nothing quite as peaceful or contemplative as sitting in the afternoon sun and watching them at play.

The brilliant blue flowers backed by silver foliage would be enough to recommend this plant, but the fact that the blooms arrive in fall and become more intense as the season wears on is sheer serendipity. As well, the silver foliage seems to hang on well after frost has nipped away at everything else.

Since it blooms on new wood, a buddleia should be cut back each spring. This will give a perpetual youth to its branches marvellous to behold. The shrub gets a little bigger each year but nothing outrageous. And, if you need to keep it in a confined space, it won't object to pruning.

This is a sun-loving plant that likes open spaces. I have one close to the deck so I can admire its intrinsic beauty as well as the myriad insects that swarm around. It is also just about the perfect screening plant. The idea is to place shrubs so that the viewer looks through plants to see what comes next rather than being terribly obvious. It's a very sensually textured way to go at making a small garden seem much larger. The buddleia close to the deck provides a veil through which we see the rest of the garden, providing a kind of resonance no other plant could achieve in this site.

This native of China is a hardy deciduous scented shrub that flowers from August to October. Now, most plants that get this kind of billing might bloom sometime during that period, but this one really does keep coming right on until frost—as long as it's consistently deadheaded. Otherwise the candle-shaped racemes of the blooms turn a dull brown and look kind of ratty.

Shove a buddleia up against a wall or fence and it will look like a

PLANTING & MAINTENANCE TIPS

❧ Hardy in zones 5 to 9, this shrub should be kept out of the wind. It isn't fussy about soil but it should be well drained. Have it close to a water source in summer. Add lots of compost or well-rotted manure to keep it fed well enough to support all the desired florescence.

❧ It will seed itself all over the place if you fail to keep it deadheaded.

climber. Keep it low and mounded, and it's a fine plant in a perennial border. Don't worry if it gets killed back in the winter; creative pruning will rescue it in spring. It will grow to 15 feet (5 m) but more normally 5' – 10' (1.5 m – 3 m).

This plant looks absolutely splendid with artemisias. The silver in each plant works well together. It would look great with any plant with deep green leaves such as chelone (turtlehead), which blooms at about the same time. I have one area in a great combination of blues and yellows: *B. d.* var. *nanhoensis* 'Petite Indigo' with *Liriope muscari* 'Variegata' and a brilliant yellow hosta.

Buddleia davidii

Buddleia davidii var. *nanhoensis* 'Petite Plum'

PRUNING TIPS

❧ Cut *B. d.* in spring to about 3 inches (7.5 cm) from the base. Each year it's enthralling to watch the miraculous rebirth.
❧ *B. alternifolia* needs to be kept under control; chop one third back after blooming. However, the less you touch it, the more spectacular the blossoms will be. In milder areas where it probably won't die back in winter, remove weak canes and cut back all the others by at least half.

OTHER SPECIES & CULTIVARS

B. alternifolia has foliage that will be almost hidden by masses of bloom. They start growing later than most other shrubs, so don't despair. Because it blooms on last year's wood, pruning means the loss of a season of flowering. Cut tips back after bloom to encourage more flowers the following year. Grows to 10' – 15' (3 m – 4.5 m) in an open sunny place; zone 4.

B. a. 'Argentea' is considered the choicest of all buddleias with grayish willow-like leaves, light green above, silver beneath. It can be easily staked and pruned into a graceful single or multitrunked small tree. It needs lots of mulch in spring.

B. davidii var. *nanhoensis* (Nanhos are the dwarf form) is generally under 4 feet (1.2 m) high and wide; 'Nanho Purple' is deep blue; 'Petite Indigo' has cobalt blossoms.

Caryopteris x *clandonensis*

FAMILY NAME: *Verbenaceae* / ZONE: 5
PHOTOGRAPHED IN THE GARDEN OF: Paddy Wales

Autumn, the most glorious of seasons, is graced by the silver foliage of caryopteris covered with bright blue blossoms gleaming in the crisp sunlight. This plant combines magically with lavenders, viburnums and heaths, which act as foils: light and airy contrasted with thick or dark green foliage.

I have it close to *Ceratostigma plumbaginoides*, a shrubby ground cover of great beauty whose deep blue flowers bloom about the same time. The culmination of the season takes place when the caryopteris turns silver, the ceratostigma a brilliant scarlet and the *Viburnum plicatum* 'Summer Snowflake' is still in bloom. It's a theatrical sight, especially since they are linked by a *Perovskia atriplicifolia*, Russian sage, which has even more

Caryopteris x *clandonensis*

silvery foliage and paler blue flowers. Mine doesn't really get enough sun to stand up straight so it flops about, but I don't mind since it's draped over *Hedera helix*, the evergreen of English ivy. My hero Russell Page, the great English garden designer, also loves this combination. I came to it on my own and it's a wonderful thrill to have echoed his taste inadvertently.

This English hybrid is a deciduous small shrub or subshrub with opposite short–stalked leaves that are silvery on the bottom. They look quite mysterious in a wind. The small flowers are borne on new growth, which means that they grow on the newest part of old stems or along the new growth of any stems you've cut back. It grows to 3' – 4' (1 m – 1.2 m).

If caryopteris suffers from winterkill, don't worry, it will revive in spring. In colder areas make sure that you leave the plant alone in fall, except for cutting off all those seedheads. Keep deadheading and you'll keep getting blooms almost to frost, although they get smaller and smaller as the season wears on. This is a profligate breeder, and you'll find babies all over the place.

Treat this marvellous shrub just like a perennial—in borders at the front or back. To keep its shape, cut it back as you would any perennial, but always in spring. That makes it a charming plant for the winter garden as well. I have three close to each other to give a small drift of ethereal color.

PLANTING & MAINTENANCE TIPS

❧ Choose a nice warm spot in the sun to bring out its best color and scent. If you have cold winters, be sure to mulch the base. It will seed all over the place. So far I've been lucky—the seedlings produce flowers that look just like the real plant. This plant will take drought and prefers light soil.

PRUNING TIPS

❧ In spring, cut back by a good two-thirds to keep a good shape. I usually vary how I prune with the three I have: one I clip lightly for shape, another I chop back very hard, the other I take back by about one-third. When the one that's clipped lightly gets too big for its apportioned space in several years, I'll cut it back hard.

❧ In areas where there is no winterkill, cutting back in spring is necessary, since it blooms on this year's wood. In warmer areas, letting them go on without pruning will mean a very large and perhaps unruly plant.

OTHER SPECIES & HYBRIDS

Caryopteris x *clandonensis* 'Kew Blue' has dark blue flowers; 'Blue Mist' is powder blue. *C. incana*, blue spirea, has gray aromatic leaves; 4' – 5' (1.2 m – 1.5 m) if left unpruned.

SUMMER-SWEET, SWEET PEPPER BUSH

Clethra alnifolia

FAMILY NAME: *Clethraceae* / ZONE: 3
PHOTOGRAPHED IN: VanDusen Botanical Garden

Clethra alnifolia

Walking through the shady part of my garden in August, I'm always swept away by the sight of a *Clethra alnifolia* in bloom. Sweetly scented flowers on a small shrub back-lit by the dappled afternoon sun create an almost perfect garden picture.

This little plant is ideal for the shady garden. It can also put up with salt air and spray, a slightly soggy site and alkaline as well as acid soil. And it can take full sunny dry sites as well. Clethra grows to about 6 feet (2 m) on average and usually no higher than 8 feet (2.5 m). It could be just the plant for someone who feels reluctant to prune a handsome shrub.

In fall the leaves turn almost yellow to bright red, but they don't stay long enough to give a real show, and the pale white or pink flowers aren't flamboyant either. Scented racemes (stalked flowers arranged singly along

Clethra alnifolia

an elongated axis) bloom from August right through to the end of September without deadheading, and for this reason alone I'd want it in my garden.

Leafing out is usually quite late, which may panic the first-time clethra owner. When it finally comes through, the foliage is a pale, almost yellow green. The leaves are dense, and the oval shape of the plant will slowly form into a clump or even a colony if all the conditions are ideal. I underplant mine with tons of spring bulbs such as species narcissus, wood hyacinths and scilla, followed by a few delicate little pale lemony hostas. Later on, lilies, lamium, liriope and hostas come up around this accommodating plant.

On a country property, clethra is the perfect plant to naturalize around the edge of a pond or small lake. I like it best in the border mixed with other shrubs such as *Enkianthus campanulatus*, small azaleas, rhodos or viburnums and surrounded by perennials.

I haven't noticed any pest bother this plant. But the literature points out clethra could be subject to mite damage in a sunny exposure, especially during prolonged dry weather—just another reason for putting it in some shade.

PRUNING TIPS

❧ Though it doesn't need regular pruning, clethra looks best if given some kind of shape to keep it from getting too twiggy. Cut out old growth while the plant is dormant in late winter. Take out any dead wood.

❧ *C. acuminata* and *C. barbinervis* can get fairly large. Take off the lower branches and cut off any growth that can't be accommodated. Do serious shaping during the cold months before the plant leafs out in spring. Mulch with pine needles or finely ground bark chips.

PLANTING & MAINTENANCE TIPS

❧ The ideal situation is light shade and lots of water. If there's going to be water deprivation, make sure you don't plant it in full sun. I always keep mine mulched on the theory that this is a mountain plant and should have cool soil. When you place the plant, take into account that it will increase by underground stems to make a pleasing clump. I throw a bit of compost at it each year and keep thinking I'll do more, which I'm sure the plant would appreciate.

❧ Native clethras such as *C. acuminata* won't tolerate much fertilizing, so just add the odd bit of compost, perhaps with a little manure, to keep it thriving.

OTHER SPECIES & CULTIVARS

C. alnifolia 'Paniculata' has long, hardy vigorous terminal panicles. Reaches 8 inches (20 cm). 'Pink Spires' opens to soft pink and fades to light pink; 'Rosea', pink buds and pink flowers fading to white, has glossier foliage than the species and blooms just slightly later.

C. acuminata, cinnamon clethra, grows taller in native habitat along the eastern seaboard, to almost 10' by 5' (3 m by 1.5 m), with shiny pale green leaves. It's a suckering plant that flowers in late July. The bark is especially suited for our long winters since its exfoliation is as interesting as anything else about the plant. Has delicate, sweet to almost spicy white florets from August to September and blooms on current wood.

DOGWOOD

Cornus kousa var. *chinensis*

FAMILY NAME: *Cornaceae* / ZONE: 5
PHOTOGRAPHED IN THE GARDENS OF: VanDusen Botanical Garden and Pamela Frost

There's no such thing as a bad dogwood, just a misplaced one. All dogwoods are wonderful and *Cornus kousa* var. *chinensis* is a superior member of an exceptional family. There are at least 40 species in the genus Cornaceae and the variety is quite astonishing, from size (creepers to giant trees) and shape, to color of leaves and twigs. The blooming season depends on the species. It's definitely a family to consider collecting, since most of them are native to northern climate zones and do well in North America.

All dogwoods flower. The showy floral parts in flowering dogwoods are petal-like involucral bracts—an involucre is a collar of bracts (tiny leaves associated with flowers) beneath a cluster of flowers. Each real, tiny flower has four sepals, petals and stamens, and one style. The drupes, plum-like fruits, which birds love so much in winter, are bright red.

Cornus kousa, a native to Japan and Korea, was imported into Europe in 1875. Unlike American flowering dogwood, *C. kousa* blooms in summer after its foliage is well out and differs in that the fruit clusters into pink conglomerations. This elegant shrub grows to 21 feet (7 m), making it almost a small tree. The white flowers are surrounded by diminutive creamy bracts with pink tips and red fruit in autumn. The pointy elliptic leaves that curl under slightly are 2" – 3" (5 cm – 7.5 cm) long. The upper part of the leaf is dark green, the lower has a hairy surface and tufts of brown hairs in axils of the main vein. The glorious scarlet autumn foliage hangs on for a long time.

C. k. var. *chinensis* is native to China, and its bracts appear slightly later than the former. It's also less demanding about the acidity of the soil than *C. kousa.* I love the shape of this plant. The long, elegant limbs with glorious shiny smooth green leaves branch out horizontally, giving it a delicate look. The opposite undivided leaves are without lobes. Their distinctness makes them easy to identify.

This very slow growing plant, good for a small city garden, may take several years before it flowers. The white bracts just above the leaves along horizontal branches fade to pink and develop bright red fruits. The fall color is spectacular.

Cornus kousa var. *chinensis*

Recommended for a woodland section of the garden along with viburnums, the broad-leaved evergreens, hostas and plenty of small spring bulbs.

PRUNING TIPS

❧ Don't prune *C. k.* or *C. k.* var. *chinensis.*

❧ In the case of *Cornus alba,* which has bright red twigs, about a third of the branches should be removed every year or every other year to keep its youthful brilliance.

PLANTING & MAINTENANCE TIPS

❧ *C. k.* var. *chinensis* likes either full sun or dappled shade in well-drained, acid to slightly alkaline soil. Doesn't like a really dry situation at all. The more sun, the more fruit in this case.

OTHER SPECIES & CULTIVARS

Cornus kousa var. *chinensis* 'Milky Way' and 'Speciosa' work well in the warmer parts of zone 6.

C. mas, European Cornelian cherry, deciduous shrubs or trees; grows to 25 feet (8 m); zone 5.

C. florida, eastern flowering dogwood, is a magnificent horizontal branching shrub native all through eastern Canada and the United States; grows to 15' – 20' (4.5 m – 6 m); bright red berry-like fruits, zone 4. There are some great hybrids.

C. f. 'Welchii', a tricolor dogwood, zone 4.

C. alba 'Sibirica' is a splendid shrub with bright red stems; good in the winter garden. Grows to 7' by 5' (2 m by 1.5 m); takes some shade. I use it to screen the compost bins, which it does very well indeed. This plant should be trimmed back, since the new young shoots are a more brilliant red than the old. The cut branches make a marvellous addition to the winter garden in big planters combined with bits of pine or cedar; zone 2.

C. a. 'Elegantissima' (syn. *C. a.* 'Sibirica Variegata') has typical dogwood leaves with white margins and a motley green inner space. It will perk up the gloomiest part of a shady garden.

C. alternifolia is a spectacular tree that will grow under the worst conditions; zone 3.

C. nuttallii, western dogwood, native along the west coast from British Columbia down to California. Its bracts appear in March to the end of April; zone 8. 'Gold Spot' ('Eddiei') is a cultivar with gold spots on the leaves. Flowers in spring and autumn.

Cornus kousa

ROCKSPRAY COTONEASTER

Cotoneaster horizontalis

FAMILY NAME: *Rosaceae* / ZONE: 5
PHOTOGRAPHED IN THE GARDEN OF: Kathy Leishman

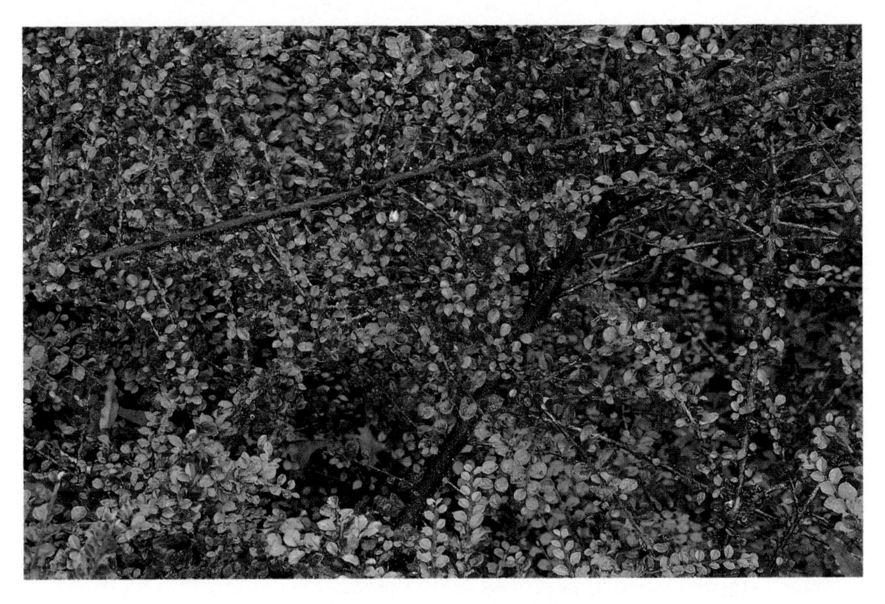

Cotoneaster horizontalis

Cotoneasters are the workhorses of any garden. The small bright green leaves, the brilliant red berries and variety of shapes and sizes make them a filler plant par excellence. They range in size from only a few inches (centimetres) to 10 feet (3 m) high and are all spring bloomers. They are tough, hardy, easy-to-grow plants that don't require much maintenance.

Several are exceptional for their fall color: *Cotoneaster simonsii, C. horizontalis, C. divaricatus, C. dielsianus, C. foveolatus.* Some are evergreen in certain regions, others are deciduous and turn bright orange-red in fall.

The pink or white flowers usually come in clusters followed by a spectacular display of red or black berries and, rarely, even yellow. The berries attract desperate birds in winter.

The long slender branches and the fountain-like arching habit of some of the larger forms make them ideal plants for the winter garden, the foliage border or the perennial garden. The low-growing forms are good

PLANTING & MAINTENANCE TIPS

❧ Most cotoneasters prefer full sun but some will accommodate part shade. They all like the excessive dampness of deep shade. Almost any moist well-drained soil will suit since they are tolerant plants and, once well established, can even take some drought. They don't transplant well, so it's best to buy container-grown plants and plant them in the right place first time out.

ground covers, with rooting systems deep enough to make superb soil binders. This is a smashing solution on a steep eroding slope. And they tolerate both exposed sites and pollution.

It's a difficult family for identification, however. Cotoneaster can set fertile seed without flowers being pollinated, which is the usual route for most plants. This is called *apomixis* and it results in all sorts of minor variants within the same species, so it's often difficult to make an exact identification.

Cotoneaster horizontalis, rockspray cotoneaster, is my favorite in this huge family. It's a low-growing shrub, and spreads out its forked branches in a herringbone branching pattern. In the northern parts of its range leaves drop off, but in warmer areas it will be semi-evergreen. The leaves are an attractive greenish gray. Grows to 2 feet (60 cm); the width of this plant is a remarkable 4' – 5' (1.2 m – 1.5 m); from zone 5 to 8.

Mine grow in two places I think are particularly attractive. One flows over the edges of a raised bed, producing a fountain-like effect. The other has been trained up against a wall. The flowers are a delicate pink and the fruit brilliant red. In autumn, the color is a short-lived orange-red.

C. h. 'Variegatus' is a hybrid similar to the species but its growth is even more sluggish. The margins are a pale creamy white without nearly as much fruit. *C. h.* 'Robustus' is a more vigorous variety than either of the others.

Under their curved branches grow dozens of little bulbs: muscari, scilla, snowdrops—all the treasures of early spring. The unfurling of leaves covers the detritus. There are small forms ideal for the rock garden. Others make a tailored ground cover. Try as many as you have room for.

PRUNING TIPS

❧ These plants require little pruning other than to take out dead branches or to keep the willowy shape. If necessary, prune in summer after flowering is over. Cut into at least 4 inches (10 cm) of healthy wood, and prune so that the stems will arch outwards. Thin overly dense plants at the base.

❧ Always remove dead wood from low-growing forms and never shear them—or you'll destroy the whole plant.

OTHER SPECIES & CULTIVARS

Cotoneaster adpressus, creeping cotoneaster, is a deciduous, very slow growing type (it could take 50 years to reach its full height of 1 foot (30 cm). Branches spread 6 feet (2 m) with pink flowers, bright red fruit and small dark green leaves; zone 4. *C. a.* var. *praecox* is a more vigorous, slightly larger form.

C. apiculatus, cranberry cotoneaster, semi-evergreen shrub 3 feet (1 m) high with round leaves; spectacular crimson to purple foliage in autumn; zone 5.

C. dammeri 'Skogsholmen' is a terrific ground cover with bright, bright green leaves; spreads to 3 feet (1 m) at less than a foot (30 cm) in height; zone 5.

C. dielsianus var. *major* is one of my favorites. The large fountain shape has almost green-gray leaves that are real show stoppers; zone 5.

C. divaricatus, spreading cotoneaster, from 3' – 7' (1 m – 2 m) with wide-spreading branches. The leaves turn orange and red in fall and the shrub is covered with brilliant red berries. It's one of the loveliest and easiest to grow; zone 5.

C. franchetii is a semi-evergreen shrub with almost sage green leaves and round orange to scarlet fruit held on long, slender arching branches; grows 6' – 8' (2 m – 2.5 m) and

Cotoneaster splendens

ENKIANTHUS

Enkianthus campanulatus

FAMILY NAME: *Ericaceae* / ZONE: 5, and parts of 4
PHOTOGRAPHED IN: University of British Columbia Botanical Gardens

Every so often a bit of serendipity comes into the placement of plants. Case in point in my own garden is an *Enkianthus campanulatus.* It's next to a *Geum* a hortbuddy gave me that is quite unlike any other I've seen. The flowers are like pale coppery pink silk. It is exactly—exactly—the same color as the pink-striped bells of the enkianthus. Gardeners call this phenomenon color echoes.

The bell-shaped flowers are indeed like a campanula, but in this plant the blooms are a mysterious pink with red streaks and hang at the end of the branches in a stunning arrangement.

Enkianthus is in the same family as rhodos and azaleas—the Ericaceaes. Native from the Himalayas to Japan, this plant is deciduous and fairly tall—about 6' – 8' (2 m – 2.5 m) generally, but in slightly warmer areas it

Enkianthus campanulatus

PLANTING & MAINTENANCE TIPS

⁂ Enkianthus prefers acid soil but does quite well in my clay soil. It likes indirect light. To give protection from too much heat, keep it out of direct midday sun. Another good plant for partial shade. But it needs pretty consistent watering and good drainage. Add peat and leaf mold to the surface.

will grow up to 15 feet (4.5 m). The leaves clutch together at the end of branches and are almost blue-green in summer. In fall they turn a thrilling vivid scarlet with plenty of blue mixed in. For that reason this is not an easy shrub to place. Next to anything with murky yellow or orange autumn tones, it could look quite drecky. It also needs cool weather to turn this spectacular color. In warmer areas the leaves may fall off the shrub while they are still green. But that's no reason not to choose it.

Another of its great virtues is that it doesn't take up a lot of lateral space. Flowers grow on the previous season's growth; buds form at the end of branches each year. The bark is green-brown. The colder the area, the more restricted its growth.

The upright shape makes this a really terrific plant to put in a mixed border. And enkianthus is a superb mixer. It goes particularly well with ornamental grasses. The harmonious vertical forms of both kinds of plants complement one another. In winter the rustling of the grasses near the shrub adds an eloquent sound to the garden. Given the fascinating upward growth of the branches, it's a good shrub to plant with the larger spring bulbs, especially narcissus, and species tulips such as *kaufmanniana* hybrids and varieties.

As a pleasant grace note, mix with different campanulas such as *Campanula persicifolia*, peach-leaved campanula, which is also a vertical plant and grows under the same light conditions.

PRUNING TIPS

⁂ Little pruning is required except to remove dead or broken branches. Snip off any short little spurs that might develop along stems of old plants. If you want more branching, cut it back after flowering. If it gets too dense at the base, cut back to stimulate budding. In older plants, cut back at least one shoot to the ground.

OTHER SPECIES & CULTIVARS

Enkianthus campanulatus has yellow-orange flowers streaked with red at the base in multi-bloom clusters; var. *albiflorus* has whitish flowers; var. *palibinii*, reddish flowers.

WITCH ALDER, AMERICAN WITCH ALDER

Fothergilla

FAMILY NAME: *Hamamelidaceae* / ZONE: 5
PHOTOGRAPHED IN: VanDusen Botanical Garden

Fothergilla has a huge reputation as a temperamental, tricky to grow plant. There's nothing like this sort of a challenge to prod the gardener into experimenting. After attempting a few of these gorgeous shrubs, I think the problem is with the gardener, not the plant.

It can take a fair whack of abuse without a murmur. For instance, I bought a tiny little one and nursed it along in absolutely the wrong circumstances (under a maple tree, which is completely unfair given all the shallow nutrient-gobbling roots it has). After a couple of years of not much

Fothergilla gardenii

PLANTING & MAINTENANCE TIPS

❧ For best results, find a site that will give the plant at least half a day of sun. The flowers will be fuller and the leaves a more pronounced color in full sun. But in half shade and sun later in the day, mine still looks quite striking and produces more flowers than one grown in deeper shade.

❧ In the dappled shade of woodland with lots of humus added to the soil, it will be rounder and more compact in habit. Plant slightly higher, about 2 inches (5 cm), than the surrounding soil and mulch well. In its native habitat it will grow in any number of soils. But keep it out of drying winds or where the soil is going to dry out regularly.

❧ Most of the literature says this plant *must* be grown in acid soil. I was really cheered to read that John Sabuco has grown it in clay, which is what I've got. And, like him, I have never had a plant suffer from chlorosis (turning yellow), and only real clumsiness has killed off a plant.

❧ *Fothergilla* doesn't need a huge amount of water after the first year and isn't bothered by pests or fungi. Even if you don't have perfect drainage, and have a bit of acid in your soil, this plant should be okay.

action, I moved it to a better spot, and the *Fothergilla* responded almost immediately. It's been growing quickly (for a *Fothergilla*) ever since. Same soil (heavy clay), same light conditions (half day of sun) and same drainage (not great). That, to me, doesn't sound like a fussy plant.

But there are no hard-and-fast rules in gardening. I planted a more mature form, changed my mind about its placement, moved it, and watched it croak almost immediately. I think the message is quite clear. This gorgeous plant is worth trying no matter what your circumstances.

The most familiar forms of *Fothergilla* are *F. gardenii* and *F. major*, and it's a good idea to make sure which one the nursery sold you and to know the difference between them. *F. major* hates having wet feet; *F. gardenii* is a swamp plant. On the whole either can cope with sun or light shade. They are among the first shrubs to start leafing out in spring.

F. gardenii reaches 2' – 3' (60 cm – 1 m) and has a strong, very compact, dense appearance. The cream-colored bottle brush flowers, about 1½ inches (3 cm), appear in early April or May (depending on your area). The buds might start to break out in March, then the blooms materialize above the about-to-unfurl leaves, which fully reveal themselves only after the flowers have fallen off. At no time are the flowers masked by the attractive leaves. The foliage is a bit like witch hazel, only smaller, in a pale gray-green that turns to medium green in summer. The transition in fall is absolutely fascinating because the leathery leaves may turn yellow, orange, green and red all on the same branch, even on the same leaf. It's called a bonfire effect and is a real eye stopper.

Plant *F. gardenii* around the edges of a pond or in boggy depressions. It is small, deciduous and has slender crooked, often spreading branches. On the whole it's useful anywhere a dense form is needed in a design. It goes well in borders, for massing or up against a house or fence. Combine it with rhodos and azaleas. The slender branching shape in winter is the perfect complement to these broad-leaved evergreens.

In the east, *Fothergilla* comes out at the same time as *Cercis canadensis*, Eastern redbud, so this becomes a natural and marvellous marriage. Given the size and shape, other dark green low-lying ground covers—such as *Hedera helix*, English ivy; *Paxistima canbyi*, ratstripper; or *Vinca minor*, myrtle—make a splendid background for the very strong foliage.

The larger form, *F. major*, may grow as high as 10 feet (3 m) and makes a really good screening plant; its dense nature can mask out unsightly areas. I have a *Fothergilla* combined with *Stachys*, lamb's ears, and a low-growing hardy geranium near the brilliant twigs of *Cornus alba* 'Sibirica'. These plants look wonderful with evergreens such as hollies, mahonia, the deep green of *Taxus*, yews, and *Tsuga*, hemlock. It blends in well with the sensuous nature of the woodland.

PRUNING TIPS

❧ About all you have to do to keep it in fine fettle is to remove any dead stems—take them out right to the ground. Any pruning is best done in winter when the plant is still in dormancy or after flowering in spring. When it suckers from the ground, remove them or plant them elsewhere. Do an occasional thinning in the centre of the shrub to let in light.

OTHER SPECIES & CULTIVARS

F. gardenii, dwarf fothergilla, forms thickets in its native habitat from Virginia to Georgia. It will grow 2' – 3' (60 cm – 1 m) in part shade; and more than 6 feet (2 m) wide in sunny sites. *F. g.* now includes *F. parvifolia*; *F. major* now includes *F. monticola*; *F. major*, large fothergilla, is native from the Allegheny mountains to the coastal plain down to Georgia. It's slightly coarser than the dwarf fothergilla and grows to 6' – 8' (2 m – 2.5 m). It is slightly hardier and flowers later, after the leaf comes out in early spring. It is hardy to –30°F (–35°C). The more robust alternate leaves are bigger with zigzag stems. The warmer the area, the more vivid the blaze of autumn color. In fact, this is the ideal plant to use where there is little or no fall color.

WITCH HAZEL

Hamamelis x *intermedia*

FAMILY NAME: *Hamamelidaceae* / ZONE: 4 through 7
PHOTOGRAPHED IN THE GARDENS OF: Dick and Barbara Phillips and the VanDusen Botanical Garden

This North American native is the divining rod of legend—settlers used the forked branches to search for water. It's found in damp woods from Nova Scotia to Nebraska and as far south as Georgia. The wonderful fall color is almost enough justification for this plant. But it's the early spring fragrance and its virtually frost-resistant flowers that are the *coup de grace.* Some bloom in fall, others bloom at the last snowfall.

Pests don't bother this plant, diseases are rare and there's not much work needed to keep it looking attractive. This makes a perfect choice for the bones of your garden if you want something hardy, reliable and well formed. It is, however, very slow growing. I have two and they are placed near other

Hamamelis x *intermedia* 'Pallida'

plants that are more glamorous during the summer: *Cimicifuga*, snake root, and *Thalictrum*, meadow rue, for instance.

This plant comes into its glory in fall and winter. The distinctive leaves have scalloped edges and the shrub itself is vase-shaped. The flowers, which bloom in late winter or early spring, are fragrant enough to stop you in your tracks during a garden stroll. And the fall color ranges from golden yellow to flame, depending on the variety. The vase shape of this shrub looks particularly striking when underplanted with bulbs flowering at the same time. I have tons of *Narcissus* 'Tête à Tête' and 'Jack Snipe' (both have small pale yellow blooms) interspersed with grape hyacinths.

One of the quirky things about this shrub is that it bears both flowers and ripe fruit at the same time. The double-chambered pods take 12 months to ripen and they are absolutely amazing. When they dry out and consequently shrink, the black pods can explode, spewing seeds some distance away.

H. x *intermedia* grows 5' – 8' (1.5 m – 2.5 m) on a single smooth brown stem. Flowers are singular in their twisty shapes, like some exotic butterfly. It is quite astonishing to see them sitting perched on a leafless stem when there is just about nothing else in bloom. This is a northern gardener's ideal plant. If it gets too cold, the flowers just curl up into themselves.

Hamamelis x *intermedia* 'Orange Beauty'; *Viburnum tinus* 'Spring Bouquet'; *H.* x *intermedia* 'Copper Beauty'

It also stands up to pollution and is hardy in zones 4 through 7. Thus it's ideal for plantings near roads and in other difficult-to-grow areas. It will easily naturalize along streambeds, in woodlands. As an inner-city plant, there are few better or tougher. And, because it grows slowly, it's ideal for a small garden.

PRUNING TIPS

❧ If you are aghast at having to do regular pruning, you'll love this plant, since it usually doesn't need pruning at all. But if pruning *is* necessary to keep the attractive vase-like shape, do it right after flowering.

PLANTING & MAINTENANCE TIPS

❧ *Hamamelis* thrives in almost any soil, from neutral to acid. It does need a fair amount of water, especially if you are growing it in heavy clay. In wet, poorly drained places use this plant instead of a rampaging one like any of the salixes (willows).

❧ Likes full sun to light shade and does best in moist soil with loads of leaf mold or organic matter added.

❧ The only pests that seem to attack it are the following, and they are rare: *Phyllosticta hamamelidis*, a fungus that causes spots or browning. Don't plant near birch or it might contract leaf gall insects (*Hornaphis hamamelidis*; birch alternative hosts).

OTHER SPECIES & HYBRIDS

H. x *intermedia* 'Arnold Promise', an American hybrid with bright yellow, fragrant flowers that blooms early in March; has yellow fall color, and grows to 25 feet (8 m), but averages 10 feet (3 m) square. If frost hits, the flowers curl up for protection; zone 5.

H. vernalis is smaller, 6' – 8' (2 m – 2.5 m); late winter red-centred flowers give a coppery glow to winter landscape. This most common witch hazel is native to the southeastern United States, with narrower leaves, yellow to red blooms. The crimson fall color can last for almost a month; zones 5 to 9.

H. mollis, Chinese witch hazel, grows to 33 feet (10 m) in its native habitat in China. Yellow flowers from January to March.

'Pallida' has pale lemon yellow flowers; other hybrids have yellow flowers suffused with red, giving an orange appearance from a distance; zones 5 to 9.

H. virginiana, a native from eastern North America, has golden yellow autumn color. This species is very showy when in bloom, though the flashy yellow leaves in autumn hide the paler yellow flowers; grows to 16 feet (5 m); zone 3.

Hydrangea quercifolia

FAMILY NAME: *Saxifragaceae* / ZONE: 5
PHOTOGRAPHED IN: University of British Columbia Botanical Gardens

Hydrangeas have gotten a bum rap from me in the past. In my neighborhood, we are inundated with particularly blowsy old mop-heads with dirty white flowers that sag and loll about close to the ground. Since no one prunes them properly, they look so, well, so very ordinary.

Well, of course, it's dead wrong to lump all hydrangeas together. *H. macrophylla* are gorgeous, show-stopping hydrangeas with glorious blooms. These are divided into two forms: lacecaps and mop-heads (*Hortensia* varieties). Both originate from Japan. The lacecaps are more delicate, with central clusters of flowers.

And then the most special hydrangeas of all—*H. quercifolia*, oak-leaf hydrangea, is a North American native. This little branched mounding, rather bushy shrub grows to 2' – 6' (60 cm – 2 m) with the same width. The shrub has panicles of white flowers in fall; they turn slowly to pink and then the brown of ancient lace.

The bark alone is worth having this shrub for. It's one of the exfoliating forms. The underbark is a dark, almost sienna brown; when it peels off it reveals next year's bark. The same red brown appears on new stems and petioles.

I love this shrub because it's perfectly happy in fairly deep shade; however, any woodland situation is fine. Among its virtues is that it is pretty much left alone by pests.

As the name implies, the leaves are shaped like oak leaves, with five or

PLANTING & MAINTENANCE TIPS

- This species of hydrangea tolerates heavier soil and won't mind most soils as long as they are moisture-retentive and well drained. You can get this effect by adding lots of humus to the surface. If you do place it in full sun or plant in clay soil, be sure to add a thick layer of mulch to the surface of the soil.
- In colder areas such as zones 5 and 6, the tops may be killed and there will be fewer flowers. Protect them in zone 5, especially for the first few years.

Hydrangea quercifolia

more lobes. They look gorgeous all summer and are so large they make this shrub seem larger than it is, and very dense. The more sun the shrub has, the more luxurious and richer the foliage color. It works perfectly with some of the more strident ornamental grasses simply because it has so much character of its own. Or with *Cimicifuga racemosa*, snakeroot, which has similarly shaped leaves and makes a splendid foil. Grows to 6' – 12' (2 m – 3.5 m). Wonderful used as a bridge between the wild display of bulbs in spring and the full flush of perennials coming into their own. It's ideal for naturalizing but can also be trained up a wall, where it will grow taller than a plant in an exposed space. In fact, the warmer the area, the better this plant performs.

PRUNING TIPS

❧ It suckers from the root in a wide, spreading clump. Replant to add to your stock or keep them pruned out. Remove any dead branches and thin out so that the branches are well spaced. Cut back to a point just above the base of the previous year's growth after flowering. It can also be pruned to the ground each spring to make a compact 3 foot (1 m) shrub.

❧ For plants that have not bloomed properly, try removing spent flowers to help force remaining buds. Cut stems just above the round pair of leaves below flower heads.

MOUNTAIN LAUREL

Kalmia latifolia

FAMILY NAME: *Ericaceae* / ZONE: 5
PHOTOGRAPHED IN: VanDusen Botanical Garden

Discovering mountain laurels was such a bonus for the woodland area and the winter garden, I wonder why it took me so long. A glorious evergreen shrub is something every northern gardener longs for.

The one that lives in my garden has been so abused I'm almost ashamed to admit it. A branch from a neighboring tree fell on it a few years ago, splitting off a major stem, which I held together with an old rag. Not surprisingly, this didn't work. But I moved it to a new spot, pruned it very, very carefully, and it has continued to grow in its own slow, inexorable manner. In the coldest part of its range, *Kalmia* will grow to 3 feet (1 m), but in milder areas it can easily hit 7 feet (2 m) or go right to 20 feet (6 m).

Kalmia latifolia 'Olympic Fire'

PLANTING & MAINTENANCE TIPS

❧ *Kalmia* prefers leaf mold, but any good soil will do. The more unfavorable the site, the more important the soil requirements become. According to John Sabuco, if you don't have the proper acid soil with lots of humus, it's wise to try the following:

❧ Plant on top of the ground and bring unmodified regular topsoil up to the top of the root ball at a 5:1 slope; top dress with ½ inch (1 cm) of alfalfa meal; mulch with 4 inches (10 cm) shredded hardwood. Keep well watered.

❧ I mulch mine with pine needles and protect in cold winters with pine branches. I don't cultivate and just leave them alone to get on with their sluggish growth. Even if there is winter damage, they seem to come back quite nicely in spring. Sabuco points out that they can be defoliated by weather that hits –20°F (–35°C), but stem kill doesn't occur until –30°F (–35°C). In either case, they will releaf in April.

❧ In dry or windy locations use a mulch of oak or beech or other acid organic material.

❧ Always remove the faded flowers to encourage reflowering the next season.

❧ Don't plant on the south side of the house where hot summer sun can fry the plant.

❧ It flowers poorly in really dense shade and loses its nice shape; get plenty of moisture to the roots in summer.

❧ Good container plant if you have a large enough pot.

This elegant plant thrives in just about anything, including wet sites in the milder areas. But the colder you are, the more carefully chosen the site must be.

K. latifolia grows in eastern North America from New Brunswick to Ontario and down the East Coast of the United States inland to Indiana. It resembles one of the larger-leaved rhododendrons and has much the same habitat: woody, moist sites in acid, or sandy, soil rich in peat and humus.

In its native habitat it grows in such diverse areas as rocky hillsides and in the acid of swamps. Because it grows slowly, it is another good plant for a small garden. And, as I've found out, it's easy to transplant. It requires delicate care, since it has a shallow, fibrous root system. Though a *Kalmia* cannot stand a lot of competition in the immediate area, it will put up with almost anything except really limy or alkaline soils.

Latifolia means broad-leaved, and these elegant lance-shaped leaves start out almost yellow-green, then age into glossy green.

It has truly remarkable buds and flowers. The buds are almost as large as the flowers. There might be as few as five and as many as 50 flowers in a cluster, which bloom on the previous season's growth. Color varies from white to pinky rose to deep carmine with purple markings, with white stamens on dark anthers. The dots and bands of the markings around each bloom fascinate the *Kalmia* collector.

One of the truly delicious things about *Kalmia* is that they gnarl with age, becoming increasingly open and giving the garden an ancient, settled look. The green or almost reddish twigs turn gray to red-brown as they get older, exfoliating in long narrow flakes.

When everything else is stripped bare in my own little woodland, I can see this lovely shrub resting near a *Pieris japonica* (see page 47). During the summer they are accompanied by the architectural plants *Petasites japonicus* and *Peltiphyllum peltatum* and masses of hostas, making a dense, cool border that I find instantly relaxing.

Though this plant is excellent for shady borders and naturalizing, it might not be the perfect plant for a country garden. If there are any cattle nibbling on the plants, they may keel over since the foliage is toxic. The leaves and tops contain the poisonous glucosides arbutin and andromedotoxin. Alas, deer aren't bothered in the least by the poison.

PRUNING TIPS

❧ Like rhodos, *Kalmia* responds beautifully to heavy pruning but seldom needs it. Cut some of the oldest stems to the ground to encourage new growth. And always do your pruning after the shrub's flowering is over.

OTHER SPECIES & CULTIVARS

K. angustifolia, sheep laurel or lambkill, is native from Newfoundland to Georgia. Grows in the company of other plants and is hardy down to zone 1. Flowers in midsummer on current year's growth. Simple opposite leaves, small and narrow; they may droop on the coldest of days. Grows to 3 feet (1 m); compact, evergreen and flowers when few other shrubs do. Fussy about soil. It can't take either excess fertility or extreme heat. Again the warning—this plant is poisonous to cattle.

K. poliifolia, bog kalmia, bog laurel, has rosy–purple flowers in spring, with narrow green leaves; a small plant 1½' – 3' (60 cm – 1 m) wide; good for a really wet site; zone 2.

MAGNOLIA

Magnolia sieboldii

FAMILY NAME: *Magnoliaceae* / ZONE: 6
PHOTOGRAPHED IN: VanDusen Botanical Garden

Magnolia sieboldii

Everyone wants a magnolia in the garden. Some forms have huge fragrant white flowers that open in early spring long before most plants have leafed out. One of my most treasured memories is of standing at the edge of a magnificent field in the south of France filled with every imaginable magnolia. I had hit a day when so many of them were in bloom it felt as though my whole body was drifting on a sea of blooming fragrance. Though magnolias look fragile, most of them are anything but.

There are 85 species of magnolia from North and Central America and Asia. It's also one of the oldest of all flowering plants. *M. sieboldii*, Japanese magnolia, grows quickly, with graceful arching branches and tender green new leaves that like a little shade. It will flower within five years, and grows in areas as disparate as Nova Scotia, Ontario, Poland and Scandinavia. A very good plant for the West Coast.

This Japanese species (formerly *M. parviflora*) has large fragrant white flowers with bright red stamens. They open up after *M. stellata*, star magnolia, and stay open into early summer. Unlike *M. stellata*, the flowers open after the leaves mature. Blooms can reach 3" – 4" (7.5 cm – 10 cm) across, with bright red centres and pure white tepals (petals); wonderfully scented. The leaves can reach 6 inches (15 cm). Grows to 13 feet (4 m) and, if allowed, will expand into massive thickets.

The thick fleshy roots emit a spicy odor if they are broken or cut. Magnolias really hate being moved, so transplant only in early spring just as buds start to swell but before the flowers bloom. In very warm areas they should be transplanted in fall. This plant is dependent for nutrition on soil fungi attaching themselves to the roots. Remove the soil and the fungi are also removed; it will suffer mightily. Needs lots of water and prefers the dappled light of a woodland garden, although it does well in sunshine on the West Coast. Magnolias combine magnificently with rhododendrons, of course, but especially *Rhododendron* 'P.J.M.', any pink azaleas and narcissus in spring. Ferns and hostas seem a natural marriage with this large shrub.

Magnolia quinquepeta

PLANTING & MAINTENANCE TIPS

❧ Transplant magnolias *only* before any new growth starts. They like fertile, well-drained acid soil in semi-shaded sites. They must not get dried out.

❧ Never pick up the leaves—leave them as mulch—add another 2 inches (10 cm).

❧ Choose a site where snow can build up and protect the root system. The plant can be damaged by late frosts and should be in fairly protected spots. But not against a south-facing wall, which may encourage them to break out too early.

❧ They don't need a lot of attention once they've been acclimatized. Every second or third year top dress with lots of manure with leaf mold added.

❧ Never cultivate around the roots—you might damage them. The roots are large, fleshy things that decay very quickly once they've been cut or bruised.

PRUNING TIPS

❧ Prune for shape immediately after flowering. Remove intercrossing branches or thin the inside to let air and light in. Cut all the way back to another branch and don't leave any stubs—this plant is susceptible to disease. On grafted forms, any shoots originating above the graft may be kept, any below should be removed. But wait until two or three shoots have grown before you take any out. The later ones will be the strongest. Thin out excess blooms for future strength. Buds for next year's blooms are usually formed by July, so don't cut anything off after then.

OTHER SPECIES & CULTIVARS

* Japanese and Korean species have magenta-purple stamens. Chinese have red stamens.

M. salicifolia, willow leaf magnolia, has small fragrant profuse flowers with elegant narrow leaves like a willow. This one is really choice; zone 6.

M. sinensis, Chinese magnolia, is a shade lover that prefers neutral to limestone and chalk soils; needs lots of humus for moisture retention. In sandy soil, add organic matter. Tepals are pure white, rounded, 3" – 4" (7.5 cm – 10 cm), with a central cone and red, crimson or plum stamens—the effect is stunning. Grows to 20 feet (9 m); zone 7 to 9.

M. stellata, star magnolia, is the best known of all magnolias. Its white fragrant flowers appear before leafing out. Late freezes and wind may hit the flowers hard. If it's sited with a southern exposure, it might open too early. Fruits earlier than other magnolias, and may take four years before flowering. Later in the season there will be a great display of scarlet seeds. This species is less tolerant of limy soil and full sun than most magnolias. Hardy to –5°F (–20°C); zones 6 to 9.

HOLLY BARBERRY, OREGON GRAPE

Mahonia aquifolium

FAMILY NAME: *Berberidaceae* / ZONE: 5
PHOTOGRAPHED IN: University of British Columbia Botanical Gardens

When people ask me what the ideal plant is to first put in the garden, I usually recommend *Mahonia.* It's a four-season wonder. The shiny green holly-like leaves (they are so similar that people often confuse the two) slowly turn maroon in winter, then for a couple of weeks in spring they look ragged and brown. For a few weeks you think the poor thing is dead, and then, all of a sudden, shiny new pale green leaves start to come back along the stems.

In spring, the more than 100 species have yellow flowers, which bloom on the previous year's terminal growth. They are mildly fragrant and to me not particularly interesting. I like the almost black fruit that comes out after flowering. Easy to grow even in difficult spots, it will naturalize in woodland conditions, enjoys the shade and will grow underneath trees or shrubs.

M. aquifolium is found in woodlands from British Columbia to California and can reach up to 10 feet (3 m) in the wild but usually grows to about 5 feet (1.5 m) in gardens. It's tough and easy to grow. I have a couple in my garden that have been moved from one place to the next with dreadful abandon. One poor thing had its roots slightly exposed to the worst winter of freeze-thaw we'd had in years and it lived to see spring. Needless to say, this plant has not reached its mature height and is slow growing. But now that I've left it alone I expect it to take off like crazy. It lives next to a dogwood, and the two are marvellous to behold in wintertime, the one with bright red twigs and the other with a combination of green and maroon leaves.

Mahonia combines well with woodland perennials, ferns, hostas; looks good in a shrub border, and if I had the space, I'd give over a large area for it to naturalize in. It also makes an excellent hedge and is good for holding back retaining banks. As a specimen in the small garden it deserves a place of honor, especially where it can easily be seen in winter.

Planting around it doesn't seem to bother this tolerant plant at all. Hostas, lots of spring bulbs and some of the other glories of the woodland live in harmony with it. One shrub I have is surrounded by various perennials, and it makes a fabulous background for lilies.

PRUNING TIPS

☙ If you have an old plant, rejuvenate it almost miraculously by chopping it back to about 4 inches (10 cm) from the ground in spring. But apart from that, *Mahonia* doesn't need a lot of pruning unless you want to tidy it up by cutting out old branches or anything that looks sad with winterkill.

☙ As the plant grows, some stems will be almost bare of leaves, with just a few at the top. Before growth starts in spring, cut those stems right back almost to the ground.

OTHER SPECIES & CULTIVARS

Mahonia aquifolium

PLANTING & MAINTENANCE TIPS

❧ Plant in either spring or fall; in colder areas, you'll give it a better start with spring planting. The soil should have a high humus content, be fairly rich and friable, and neither wet nor too dry, though mine have all survived both drought and flooding. Be sure to mulch with leaf mold, compost and manure; keep them in spots where they won't be subject to sun scald or drying winds in winter—in other words, a protected spot.

❧ The problem I had with one of mine was dreadful drainage. The leaves kept dropping off. Once it was moved to higher land with bright shade in a protected place, it thrived.

❧ If you want a bushier plant, pinch out new growth after it has flowered in summer.

Mahonia aquifolium 'Atropurpurea' has splendid red-purple winter foliage; glaucous blue berries; grows 3' – 4' (1 m – 1.2 m) high with 4' – 5' (1.2 m – 1.5 m) spread.

'Moseri' is striking but hard to find. The coppery red young leaves turn orange then yellow before they turn green; grows 4' – 5' (1.2 m – 1.5 m) high, 3' – 4' (1 m – 1.2 m) wide.

'Compacta'; tidy; grows 2 feet (60 cm) high.

'Mayhan'; a small variety; may come true from seed; best in shade, especially in warm areas. Prune to the ground to thicken.

'Golden Abundance'; tons of glossy green leaves with red in the middle; grows 5' – 6' (1.5 – 2 m).

M. bealei, leatherleaf mahonia; compound stiff leaves up to 16 inches (40 cm) long; very cold resistant; grows to 12 feet (3.5 m) on upright stem; zone 6.

M. repens, creeping mahonia; dull blue-green foliage, black fruit in summer; spreads by underground runners; very cold resistant; grows to 10 inches (25 cm); zone 5.

M. lomariifolia, Chinese holly grape; yellow flowers bloom in late winter followed by berries; deeply divided leaves may grow to 2 feet (60 cm), with dark green spiny margins; grows to 10 feet (3 m); tender, zone 6.

LILY-OF-THE-VALLEY SHRUB

Pieris japonica

FAMILY NAME: *Ericaceae* / ZONE: 4 to 8
PHOTOGRAPHED IN: VanDusen Botanical Garden

Most gardens need the refreshing sight of a broad-leaved evergreen in the winter. *Pieris japonica* is not only a shade plant, it's a plant for the winter garden.

In warm coastal areas this plant can get too far out of bounds for my taste. They grow to be too big and blowsy when left to their own devices, a totally unnecessary state since they take to pruning beautifully. And in spring their shower of small cup-shaped white flowers is utterly delightful.

New leaves unfurl with a lovely slightly bronze tint. And the glossy lance-shaped leaves grow on the shrub in a mound-shaped form. The dark green foliage is reddish when it opens up. A sturdy, tough plant, it's especially fine for northern gardens since it has fascinating color all winter long.

Pieris japonica 'Valley Fire'

PLANTING & MAINTENANCE TIPS

- Plant pieris in early spring or in fall from September to October. It is an ericaceous plant and therefore likes the soil of the woodland—slightly acid with lots of organic matter such as compost and leaf mold; or add lots of humus to the soil in the form of pine needles or oak leaf mold.
- The warmer the area, the more important drainage is. In warm areas, plant it on a slightly mounded rise of shredded pine bark to provide this drainage. Make sure to plant at exactly the same depth as the container it came in. Water well the first year; after that during dry spells.
- In sunnier spots the flowers will be more prolific. But keep it away from the path of any strong winds. The farther north you garden, the more protected the location should be.
- The plant might be hit with chlorosis, a slight yellowing of the leaves. Add some form of chelated iron.
- Don't cultivate around it or you will disturb the shallow roots.
- Mulching in fall is an absolute must. Besides, it performs as a handsome backdrop for the plant.
- Pests: lacebugs might attack plants growing in the sun. An eighth of an inch long, they look like specks on the underside of the leaves where they are busy sucking the sap. Spray with insecticidal soap.

Flower buds dangle in clusters all winter; the snowy white, sometimes pale pink flowers resemble lily-of-the-valley (thus the common name of the plant). It has a slight fragrance. Another virtue—deer don't like to eat it, so consider it for the country garden.

I've moved my plant several times—putting it in a place too small one time, in the path of a flood another (hates wet feet) and finally settled it in a berm and close to a *Kalmia*, and they look extremely handsome in partnership. The strong shiny green leaves complement one another. The buds appear red in winter. Since I've left it alone it has been thriving. Grows 3' – 10' (1 m – 3 m).

The hybrid *P. j.* 'Mountain Fire', has strong reddish bronze new growth, and the new leaves display themselves in this form for several weeks. The white flowers aren't all that showy, which is fine with me. I'm particularly fond of the shape of the leaves and the form of this shrub.

PRUNING TIPS

❧ Though you can safely leave pieris alone, it will grow fairly tall if not cut back a little. Deadhead regularly and pinch out new growth in an overly leggy plant. To encourage branching, prune after flowering. The flowers bloom on old wood. With very old or overgrown plants, thin gradually at the base or take a chance and cut right back to the ground in early spring.

Pieris japonica 'Valley Fire'

OTHER SPECIES & CULTIVARS

Pieris japonica 'Red Mill'; bright red new growth; grows to at least 12 feet (3.5 m); 'Variegata', green leaves edged with white; more compact, dense and slow growing. Grows 5' – 7' (1.5 m – 2 m); 'Mountain Fire', heavy bloomer with white flowers and bright red new growth; grows to 10 feet (3 m).

P. floribunda, mountain pieris, a native of the Northeast; evergreen; more lime tolerant than *P. japonica*; will take full sun; serrated leathery leaves; the upright flowers look a bit like blueberry and bloom from late March to early May, depending on how warm your area is. A small plant that normally grows to about 4 feet (1.2 m).

AZALEA, RHODODENDRON

Rhododendron yakusimanum

FAMILY NAME: *Ericaceae* / ZONE: 5
PHOTOGRAPHED IN THE GARDEN OF: Joan Rich

No matter where you live, rhododendrons have become *the* favorite shrub. Rhodos or rhodies are the affectionate names that rhodophiles use for their favorite plants. These glorious shrubs are considered difficult—usually by those who know nothing about them. There are rhodos to suit almost any region in the country, a color and size to fit every garden. But you have to be very, very careful in your choices. These plants are originally from high, cold, windy places. This suggests they will be hardy plants, which they are. But these areas have particularly good drainage, usually a bit of acid in the soil and the protection of a high forest canopy.

One of the main problems with rhodos is not how fussy they are but the eyeball-searing colors of the blooms. Think of blossom combinations first, then look to the foliage. This is the reverse of how I usually look at plants, because most of the time the foliage is more important than the color of the flowers. But with rhodos some of the hybrid shades are almost neon in tone—not recommended.

Another caveat: these forest plants should never be plunked at the corner of a house where winds whip about unabated, which is where they are usually found. It's sad. Neither can they hold up to intense March sunshine, which will burn the life right out of the large leaves.

Most rhododendrons need a little protection. The rule of thumb is to protect them for at least four years until acclimatization is complete by surrounding them with open burlap tents. Then let them survive on their own.

Rhodos come in almost any size, from a few inches (centimetres) to giants 30 feet (9 m) tall. They may have exquisite little leaves or huge ones 2 feet (60 cm) long. There are about 900 species, with more being discovered all the time. And the thousands of hybrids make them almost terrifying to think about keeping up with.

The language of rhododendrons is quite special. Here are a few words you might like to know if you get more than one type of plant.
Indumentum: the light, very fine hairy surface on the bottom of leaves.

Rhododendron yakusimanum 'Exbury'

Lepidote: sun- and wind-tolerant rhododendron.
Elepidote: more fragile, must be more sheltered.

The one I've chosen here is *R. yakusimanum.* Yaks, as they are called, are lepidotes. These magnificent shrubs are from the island of Yaku-shima, Japan. The glorious bell-shaped flowers start off as pale pink buds and open out to white blooms in late spring. The leaves are almost gracious in the way they back up the flowers—glossy green, curving at the edges with a rich brown indumentum beneath. A real touchy-feely slow-growing plant. It may take 20 years to reach 3' by 5' (1 m by 1.5 m).

An excellent plant for most northern gardens, it combines particularly well with heathers and conifers with its small mounding shape. If you don't have acid soil but yearn for this plant, try it in a container at least 2 feet (60 cm) deep and about the same across. Fill with acid soil and add plenty of peaty mulch to the surface. Keep away from hot spots and in a cool moist area. The colder the region, the more important it is to look for species and varieties and avoid the more finicky hybrids.

Rhododendron yakusimanum 'Exbury'

PRUNING TIPS

❧ Always use clean secateurs when you prune these plants. Strive for a beautiful shape and never take out more than a few branches at a time.
❧ Remove any dead branches in spring, but don't touch the rest of the plant until after the bloom period is over.

OTHER SPECIES & CULTIVARS

In the rhododendron family more than with any other flowering shrub, it's important to start with small plants. This gives them a much better chance to adjust to your microclimate. Think regionally when you make your choices.

West Coast: Look for rhodos from Oregon and Asia.
R. macrophyllum, native to the West Coast; starts out with red buds that evolve into purple flowers by June.

East Coast: Eastern North American and European natives.
R. arborescens, white flowers in June; grows to 6 feet (2 m). *R. maximum,* the best of the hardy eastern natives; rose-purple blooms; grows to 30 feet (9 m). *R. carolinianum*; rose-purple flowers in May; zone 6.

Central North America: Experiment with the hardiest breeds in your area. Don't try to fool around because these plants are expensive. The colder your area, the wiser you

PLANTING & MAINTENANCE TIPS

❧ Soil and drainage are everything to rhododendrons. They are woodland plants that thrive in dappled shade away from the fierce blasts of winter. Humus-rich well drained soil is important. It is a good idea to get your soil tested to make sure there is the right amount of acid in it to keep these plants thriving.

❧ If you have neutral soil, make raised beds. Dig in well-soaked sphagnum peat moss, sandy loam, pine needles, ground-up bark mulch; spread it over an area three times the spread of the mature plant. Let it settle well before planting.

❧ Never plant rhodos too deeply. The root system is delicate and wide spreading. It doesn't send down deep roots and it won't stand a lot of competition.

❧ Build a small mound of good soil and make a depression in the top. Put the rhodo in this and spread the roots gently in all directions. Cover with soil and water. Add more soil and water again. Add about 4 inches (10 cm) of mulch. Ground-up oak leaves or plain pine needles are ideal.

❧ If your rhodo starts looking a bit yellow, that indicates the soil probably hasn't enough acid content. Add a hit of chelated iron to acidify alkaline soil.

❧ Disbud your rhododendrons immediately after they've bloomed. This is crucial to maintaining a healthy plant. It's an interesting task because you don't want to remove next year's buds (the closed tight ones next to the rather loose open sprongs that were this year's blooms). Snap the spent blooms with a quick twist of the wrist away from the new bud.

❧ Squirrels and other rodents can attack rhodos when they form new buds. Sprinkle some cayenne pepper on the bud and surrounding area.

❧ If your plant won't flower, in spring give it a good nitrogen feed of blood meal or any of the organic fertilizers specially formulated for rhodos.

❧ Rhodos attract vine weevils. Get rid of this night crawler by making a collar for each plant out of a cardboard circle with a split into the centre. Cover it with Tanglefoot (a commercial but organic gooey substance) to trap the bugs.

are to go with small, almost ground-hugging species such as *R. camtschaticum glandulosum* and *R. canadense.* Both grow as far north as zone 2.

Others I particularly like: *R. impeditum,* a dwarf form with deep blue-purple flowers that almost smother the plant in spring; zone 5; and *R.* 'Ramapo', another small, mounded, very tough species with purple flowers and steely blue-gray leaves; zone 5.

REDLEAF ROSE

Rosa rubrifolia syn. *R. glauca*

FAMILY NAME: *Rosaceae* / ZONE: 2
PHOTOGRAPHED IN THE GARDEN OF: Kathy Leishman

A hard-working plant suitable for any garden. Perfect for those who are convinced that they cannot grow roses but want to. Good for those who need a foliage plant as a foil for lower shrubs and spiky perennials. And something that has four seasons of interest.

Grow in areas as cold as zone 2 under just about any conditions. This makes it invaluable in most parts of the continent except for very warm areas. Originally from central and eastern Europe it grows 6' by 6' (2 m by 2 m).

What is particularly attractive is the foliage: small pointed red leaves with a gray-blue cast. Pretty, simple blooms are pink with golden stamens in the centre; bright red, very large hips. The shape of the branches is particularly felicitous—arching branches so supple it's possible to bend them to your own will, Japanese style. You have only to attach a stone or pot to the end of the branch to get a more cascading form.

This is a terrific plant to use as a screen. It is light enough to see through but striking enough to stop the eye. I've combined it with a glorious grass, *Andropogon virginicus*, which has an interesting pink tone in autumn in its long slender stalks; and a *Cimicifuga ramosa* 'Atropurpurea'—the large palmate purple leaves act as a foil for the diminutive leaves of the rose. It

PLANTING & MAINTENANCE TIPS

- This undemanding plant grows in almost any well-drained soil. It needs a minimum of six hours of sunlight to bring out the glories of its foliage. It can form a benign hedge or a massed planting. Add compost and manure in combination around the plant in spring and then again after the first deep frost.
- Most of the literature says that no pest hits this plant. But an exception I've found is the rose midge. Overnight these little gray bugs will be creating cocoons for the adult caterpillar. I handpick the whole shrub without any resentment I love it so much. The midges usually don't do a lot of harm if you catch them at the right time in spring just before the blooms begin. This is when to be vigilant. Also watch for Rosy Wasp Gall, which looks like a pink woolly mass.

Rosa glauca

marries well with artemisias such as 'Powis Castle' or *Artemisia lactiflora.* And *Thalictrum* 'Thunder Cloud' has the same purple to blue haze on the foliage. It's an easy plant to use as a transition from one color to another.

The great thing about this plant is that it is easily available, and it seeds like crazy, if allowed. If you do find seedlings, dig them up and put them somewhere else. They tend to flower by about the third year. The thorny new growth makes an excellent hedge that should keep animals and children from hell out of your garden. It's a good plant to combine with Japanese maples, and plants such as *Cotinus coggygria,* purple smoke bush, that have the same tints in the foliage. This would make a wonderful soft drift of color in a large garden.

PRUNING TIPS

❧ You might have to remove seedlings if the birds don't get all the hips during the winter. I would love to have some seedlings, but mine doesn't seem to have particularly good hips. The plant may need some thinning, but don't cut the whole shrub back—cut back shoots, remove old canes and weak or damaged wood. Shape narrower at the top to make a hedge.

SNOWBERRY

Symphoricarpos albus var. *laevigatus*

FAMILY NAME: *Caprifoliaceae* / ZONE: 3
PHOTOGRAPHED: Wild, along a roadside on Salt Spring Island, British Columbia.

Here is the ideal shrub for someone with a lot of space they want to fill up quickly and beautifully or who wants a shrub that will work just fine in the shade. The great joy of this plant is in catching a glimpse of the brilliant white berries standing out against the midwinter nudity of the rest of the shrub. And it is the fruit that's the attraction rather than the insignificant flowers.

There are 18 species endemic to North America. You can find them in mountainous terrain, along the seashore and in the woodland—looking a

Symphoricarpos albus

PLANTING & MAINTENANCE TIPS

❧ This shrub transplants so easily it will cause no trouble whatsoever, even if you do it well into the winter months. It will cope with dry soil. In native habitats it lives on limestone and clay. Tolerates light conditions from sun through to shade, except for variegated forms, which must have sun. Mulch with compost leaf mold or organic matter.

bit like a sophisticated honeysuckle (they are in the same family), and very twiggy with lots of branches.

The leaves are charming: opposite, short, undivided, with a slight blue color mixed into the green of the foliage. The bell-shaped flowers bloom in clusters in late spring to early summer. The almost marble-like fruits come in shades from white through to coral.

Its shape and berries and its compatibility with other plants make it a four-season plant. However, because the form shown here suckers so easily, it is a double-edged sword. The downside is that it's highly invasive, so it should be considered very carefully before it's installed in a small garden. On the other hand, it will fill in spaces quickly for good ground cover or to hold up a slippery slope or retaining bank—and is invaluable in a large landscaping scheme. Consider the upright twiggy branching form for an informal hedge. You can train it into a hedge by planting suckers at least every 2 feet (60 cm). It's terrific in front of dark evergreens and makes the best of bad situations.

PRUNING TIPS

❧ Prune in March or very early spring so that the current growth can produce flowers. Take out any shoots that have become crowded or weak. Remove all dead shoots and suckers every couple of years. Eliminate about one-third of the old wood once the plant has become established. It will grow quickly at first and then slow down with age. A very old plant can be chopped right back almost to the ground and will bounce back with youthful vigor.

OTHER SPECIES & CULTIVARS

S. albus (syn. *S. racemosus*); native to rocky soil from Michigan to Quebec, as well as British Columbia; blunt elliptic leaves up to 1 inch (2.5 cm) long with pubescent (having soft downy hairs) on undersides; bell-shaped pinkish flowers ½ inch (1 cm) long; dead white to slightly creamy white fruits grow in large clusters in fall and remain throughout winter; grows 3 feet (1 m) high.

S. occidentalis; native from British Columbia to New Mexico and east to Michigan and Illinois; extremely hardy; funnel-shaped pink flowers.

Symphoricarpos albus with rosehips

S. orbiculatus (syn. *S. vulgaris*), coralberry; native of eastern North America; purplish red fruit ripens in September; leaves are light green on top and bluish beneath; very suckering; grows to 3' – 4' (1 m – 1.2 m) and often to 6 feet (1 m).

'Variegatus' often doesn't fruit; may revert to green form, so be sure to prune out anything that looks pure green.

S. x *chenaultii* is a neat, restrained plant with large red berries that grow on the top side of the branches. Marvellous for slopes if you are worried about erosion. Although it doesn't like wet winter locations, is excellent for harsh climates. Pink flowers, large white fruit; reaches 3 feet (1 m). 'Hancock'; low mass or ground cover shrub; if you aren't crazy about cotoneasters, this shrub will please you, since it's hardier, showier and more disease resistant; bright green foliage; pink flowers; in autumn, purple-red fruit; grows upwards by about 2 inches (5 cm) a year; and about 12 inches (30 cm) horizontally.

S. x *doorenbossii*, very drought tolerant; zone 3. 'Erect', a more narrow form with fuchsia to magenta berries; ideal for a hedge; grows to 6 feet (2 m). 'Magic Berry', dark pink fruit; showy against the dark background of evergreens in the winter garden; grows wider than tall, reaches 2 feet (60 cm) high; good for limited spaces. 'Mother of Pearl', huge white berries with a pink blush; grows 2' – 3' (60 cm – 1 m) high. 'White Hedge', pure white berries, stiff shrub, erect branches at a 60– to 90–degree angle to the ground; this is a non-suckering clone.

SNOWBALL

Viburnum plicatum
'Mariesii'

FAMILY NAME: *Caprifoliaceae* / ZONE: 5
PHOTOGRAPHED IN: VanDusen Botanical Garden

Most gardeners have an overriding passion. Mine is viburnums. I'd love to have every one, but in my garden I make do with six varieties and hope to squeeze in a few more as the garden evolves.

The foliage in the more than 120 species (and dozens of cultivars) is varied in both color and texture, the scent in some species is magnificent, the autumn color can be breathtaking, the flowers are glorious. Some viburnums are evergreen, and some can take shade. Viburnums are almost every size. Some species bear fruit and flowers simultaneously and bloom for months on end. And they're easy to care for. With such a paragon of a plant, what else could a collector wish for?

The leaves are opposite, and either simple or lobed. The flowers are quite distinctive, usually white (corolla five-lobed, bell-shaped with five stamens). Red, blue or black fruit, depending on the species. Some have both a sterile flower (with no stamen or pistil) to attract insects and a smaller flower that is fertile—it produces seeds—and decorative.

Their value in designating the bones of any garden, let alone defining a foliage border, is incalculable. Partly because of the different sizes, partly because of the form of this shrub, they should always be considered no matter what area you live in. As a bonus, they stand up to pollution.

Viburnum plicatum is one of the most elegant of all the viburnums. The branches slowly flatten out almost level with the ground. *V. p.* 'Shasta', which is the one I've observed most closely, takes a few years before it performs up to scratch, and then the spring show is breathtaking. It is smothered in creamy white flowers all along its branches. In my garden it is next to a pale yellow *Rosa hugonis.* They are stunning.

The one shown here, *V. p.* 'Mariesii', has simple opposite serrated leaves, that are a soft gray-green above and slightly paler below. The older the branches get, the darker gray or brown they become. Buds come out before the leaves and the blooms. In the fall this plant turns brilliant scarlet. For this reason, be careful where you place it. Grows 8' – 10' (2.5 m – 3 m) and just slightly wider, branching in horizontal tiers. The flowers, alas, have no fragrance, but that's more than made up for by the effulgent blooms. Insects don't come near these plants on gloomy days, so the amount of fruit depends on sunny weather.

'Summer Snowflake' is a splendid plant. The wild form of this plant is native to China and Japan and was collected in Japan and hybridized at U.B.C. This particular cultivar will start blooming when it's very small and keep going right through summer into fall. The flower clusters are from 2" – 2½" (5 cm – 7 cm) across, surrounded by larger sterile flowers; grows to 7 feet (2 m). Mine is responding really well to pruning that is keeping it in scale with its surroundings. It looks great in fall with the old flowers turning a tattered brown lace against brilliant red leaves; zone 6.

The evergreen species are mostly semi-evergreen in northern gardens. That is, their wonderful deep green leathery leaves hang on until late in winter, when they look sad and ratty. But they are among the first in spring to perk up with new leaves sprouting along stems. Winter-flowering viburnums should be given lots of sun. And a good rule of thumb with this species is to group together those that are doing something at the same time.

Viburnum plicatum 'Mariesii'

PLANTING & MAINTENANCE TIPS

❧ On the whole, viburnums transplant well, and though most of the literature points out that they hate clay soil, mine seem to thrive on it—though they grow very slowly. Most need a moist, well-drained soil and should be moved balled-and-burlapped or container-grown; only very small ones should arrive bareroot.

❧ Most like a site in partial shade in cool, moist soil with plenty of humus. None of them particularly like drought. Evergreen forms need lots of sun.

❧ The following species don't like really cold winds: *V. tinus*, *V. davidii* and *V. macrocephalum*. Check which type of viburnum you've purchased. Some *must* have woodland shade and moist soil with lots of humus. Good drainage is necessary when they are not in this situation. The following will take sun: *V.* x *burkwoodii*, *V. carlesii*, *V. rhytidophyllum*, *V. tomentosum*.

PRUNING TIPS

❧ No regular pruning is needed except to get rid of dead or weak branches and to maintain a good shape. The older it gets, the more gnarled the trunk will become. It's a good idea to cut out the oldest stems at the base. Keep in mind that flowers bloom on the previous season's growth.

OTHER SPECIES & CULTIVARS

V. x *rhytidophylloides* 'Alleghany'; semi-evergreen in zone 5. I love the leathery dark green leaves. Grows 7' – 12' (2 m – 3.5 m), with the major spurt in the first five years; zone 3 (where it is deciduous).

V. x *burkwoodii*; very fragrant; light pink buds open to white flowers; vulnerable to bacterial leaf spot and powdery mildew; grows 7 feet (2 m) with the same spread; zone 3.

'Mohawk', dark red flower buds opening to white petals; strong scent; compact form; resistant to bacterial leaf spot and powdery mildew; grows to 7 feet (2 m) with the same spread.

V. farreri (syn. *V. fragrans*), white flowers from rose red buds on leafless bright brown branches with ragged bark in mild areas around Christmas, in colder areas in early spring.

V. x *bodnantense*, highly fragrant; frost resistant; straight, stiff branches. Good for cutting branches to force indoors in a cool, bright place. In really cold regions, zone 2, try *V. trilobum*, American cranberry bush; it grows anywhere anyhow in the east; white sterile flowers. Hybrids such as 'Compactum' have leaves right down to the ground but don't produce fruit. Great hedge that never needs pruning.

V. dentatum, arrowwood: incredibly hardy eastern native; flowers in June; long shiny green leaves; good fall color, depending on the plant, from red through to yellow; very blue fruit; good hedging plant; in moist areas forms great thickets; grows to 10 feet (3 m).

Bibliography

Bloom, Alan & Adrian. *Blooms of Bressingham Garden Plants.* London: Harper Collins, 1992.

Brickell, Christopher. *The American Horticultural Society Encyclopedia of Garden Plants.* New York: Macmillan, 1989.

Cole, Trevor. *Woody Plant Source List.* Ottawa: Agriculture Canada, 1987.

Dirr, Michael A. *Manual of Woody Landscape Plants.* 3rd ed. Champagne, Ill.: Stipes Publishing, 1983.

Hillier Manual of Trees & Shrubs. 6th ed. Melksham, Wiltshire: David & Charles, 1992.

Hortus Third. New York: Macmillan, 1976.

Kruckeberg, Arthur. *Gardening with Native Plants of the Pacific Northwest.* Vancouver: Douglas & McIntyre, 1982.

Page, Russell. *The Education of a Gardener.* Harmondsworth, Middlesex: Penguin, 1983.

Phillips, Roger, and Martyn Rix. *The Random House Book of Shrubs.* New York: Random House, 1989.

Sabuco, John J. *The Best of the Hardiest.* Flossmoor, Ill.: Plantsmen's Publications, 1990.

Shigo, Alex L. *A New Tree Biology.* Durham, N.H.: Shigo and Trees, 1991.

Straley, Gerald B. *Trees of Vancouver.* Vancouver: UBC Press, 1992.

Street, John. *Rhododendrons.* Chester, Conn.: Globe Pequot Press, 1987.

Index